THE
NEUROAFFECTIVE
PICTURE BOOK 2

Marianne Bentzen and Susan Hart

THE NEUROAFFECTIVE PICTURE BOOK 2

Socialization and personality

Illustrated by Kim Hagen Jensen

NAP Books

TABLE OF CONTENTS

Preface

A picture says more than a thousand words.

Most of the input that the brain receives is unconscious and nonverbal, and our nonverbal experiences are what give our life its emotional flavor and quality. It is from nonverbal sensations that vitality flows into the words, images and concepts that adults are so often fixated on. This book is a picture book because pictures can bring us closer to a basic nonverbal vitality.

In this volume, *The Neuroaffective Picture Book 2 – Identity and Socialization*, we outline the twenty-year-long development of identity and interaction skills that unfolds during childhood and youth. We offer a model of five inherent motivation systems that mature during this lengthy period and explain how they are shaped by our countless exchanges with others – especially peers – as we grow up.

The previous volume, *The Neuroaffective Picture Book*, looked at the first few years of life and the interactions with caregivers that are necessary for children to develop healthy adult functionality at three levels of the human brain and personality: *sensory, emotional* and *mentalizing* functioning. It also offered an introduction to our personality model, *the neuroaffective compasses*. The development described in the first book lays the necessary foundation for the healthy development of identity and socialization described in this second book. So, if you like to have things in their proper order, you might want to read the first picture book before you read this one, but it is not strictly necessary.

Writing picture books is fun! We had a lot of fun writing this one and brainstorming ideas for Kim Hagen Jensen's lively and playful illustrations.

Enjoy!
Susan Hart and Marianne Bentzen

INTRODUCTION
Evolution and culture

In evolutionary terms, human beings are hyper-social animals. We band together in groups with other members of our species, establishing interaction patterns and behavioral norms driven by social motivation systems that we share with all other human beings. Actually, we have four motivation systems in common with all social mammals, namely *attachment; play and cooperation; hierarchy and status* and *gender identity*. These motivation systems are also found in many species of fish, reptiles and birds, so it is highly likely that they have been a part of our heritage since our primeval ancestors, the earliest vertebrates, roamed the oceans.

600 million years ago. The earliest neural networks arose in fungi and jellyfish around the same time as the ozone layer was forming.

350-400 million years ago. The brains of the earliest amphibians already had the same triune structure that characterizes the human brain today.

200 million years ago. In the earliest mammals, the limbic system and the cortex, which enable complex social behavior, were far more evolved. In modern humans, these areas of the triune brain structure still mature slowly, and in a sequence that mirrors their evolution.

550 million years ago. Left-right symmetry and photosensitive "eyes" have been found in fossilized flatworms. The flatworm is an ancestor of the earliest fish, which developed brain structures and a spinal cord some 50 million years later.

200,000 years ago. The cortex in the modern human brain has countless nerve fibers organized into stunningly complex information networks. It contains twice as many neurons as the entire brain of our closest relative, the chimpanzee.

Only the fifth motivation system, *mentalization*, appears to be significantly different in humans than it is in other social animals. Mentalization is the capacity for insight into oneself and others, and to develop that skill we first have to learn to connect emotions and complex thought processes. Mentalization includes both the emotional ability to empathize with ourselves and others and a cognitive understanding that different people have different emotions and points of view, which may vary over time.

Attachment, play and cooperation, hierarchy and status and gender interactions are easily recognized in the group behavior of our closest relatives, chimpanzees and bonobos. It is also clear that individuals of these species monitor each other's behavior and interactions. For instance, the innate social motivation systems of a baby chimpanzee are shaped by the group environment the chimp matures in. Some chimpanzee groups primarily use aggression and power to handle conflicts, while other groups tend to rely more on negotiation and emotional mediation. Group leaders have different styles, and if the group does not rebel against them, leaders and high-status individuals establish the norms for acceptable behavior. There are different norms for different subgroups – adults and juveniles, males and females, different status levels in the group – and for the perks that close friends and family can expect. Naturally, low-status group members will try to find the social behaviors and loopholes that give them the best chance of beating the system and achieving privileges above their station. Each group has its own unique culture, shaped by interactions between the forces of nature and nurture in that particular group. Young chimpanzees are born into and shaped by the group culture, and they go on to perpetuate their unique group culture over generations. When chimpanzees move from one group to another, something they typically do as young adults, they need to adapt to the culture of the new group in order to be accepted.

Chimpanzees live in extended family groups with different subgroups. Each group has its own cultural norms.

Human groups are very similar. It is easy to observe attachment interactions, play and cooperation, status hierarchies and gender interactions in human groups. We too monitor the interactions of other group members, and we develop distinct social cultures in different groups and subgroups. Our societies are shaped by traditions and by our current leaders. However, in modern societies we live in much larger and more complex groups than any chimpanzee, so even young children are often exposed to several different group contexts and cultures in the course of an average day.

Human beings also have a mental dimension that is far more evolved than that of chimpanzees and bonobos. We have developed intricate spoken and written languages for social communication, and we continuously create metaphors and stories about our world. It is this mental dimension we rely on when we mentalize, that is, when we develop empathic insight into ourselves, other human beings and animals.

Human communities also have different group cultures, and many groups contain several different subcultures.

CHAPTER 1

Personality and brain development during childhood and youth

Is our personality determined by nature or nurture?

For decades, there has been an intense debate about the relative influence of nature and nurture in the shaping of our personalities. Is it our genes or our childhood experiences that determine what kind of people we become? The answer from science is a resounding *yes*.

Nature + nurture = personality

During the later decades of the 20th century, researchers discovered that the quality of caregiving during the first few years of life had a crucial impact on personality development. This led some to theorize that those years were the only ones that really counted in shaping our brain, body and personality. However, during the 2000s, brain imaging studies of teenagers showed that both brain and personality continue to undergo significant changes into our 20s. In fact, brain maturation and the corresponding personality maturation continue, at a slower pace, until we are around 30 years old.

Personality development from 2 to 20 years of age

Our personality develops in the interaction between our biological makeup and exchanges with our surroundings. The scientific term for this process is *epigenetics*; it deals with how experiences change the way our genes are expressed, as new potentials open in some directions while other avenues are closed off. According to developmental and evolutionary psychology, our social motivation systems are biologically inherited and are then shaped epigenetically by our experiences with adults and other children during childhood and youth.

The social motivation systems begin to take shape in interactions with caregivers during our first two years of life, and it is from these experiences that we develop our first sense of how to handle other relationships in life. From around the age of 2 years, the child begins to test these acquired habits of interaction with other children. This provides the child with the new and challenging experience of creating satisfactory peer interactions, and over time, adults increasingly take a background role. Accordingly, the future adult personality is shaped by the interactional environment that the child grew up in. In the following pages, as we describe how various emotional and personal skills mature, we will assume that the developing child experienced both the support and the challenges that make these levels of development possible.

Below is a quick overview of emerging skills related to the social motivation systems during childhood and youth.

We can enjoy doing things together, or we can enjoy simply being together.

Attachment processes enable the young child to form new, lasting friendships and, later, romantic relationships.

Play can involve a shared story or project …   … and develops both our practical skills and our social intelligence.

Play and cooperation with good friends can help children make new friends and enter peer groups, where they can expand their play skills in new and exciting ways.

Status can be determined by leadership skills …   … as well as by social norms within our group culture.

Even toddlers are members of several groups with differing norms for *hierarchy* and *status*.

The norms of your subculture dictate what is desirable, important and cool …

As part of the development of *gender identity* during the teenage years, young people are confronted both with their own profound transformation process and with all the demands, norms and differences that exist in both traditional and modern cultures.

… but fortunately, you can learn to look beyond status and subculture and develop a deeper understanding of each other.

The *mentalization capacity* develops as children begin to be curious about their own and others' emotions. Gradually, they learn to recognize emotions without being completely swallowed up by them, and they develop the ability to empathize with others and with themselves. They also develop the capacity to reflect on the causes and effects of thoughts and emotions.

During their late teenage years and into early adulthood, young people improve their cognitive capacity for differentiating and reality testing. They develop long-term planning skills, and their ideas about the future become more sophisticated. Meanwhile, their improved mentalization capacity gives them a spontaneous grasp of a much wider range of nuances in interpersonal relationships. This gradually develops their leadership abilities, enabling them to delegate appropriate tasks to others.

A good team leader has to have a sense of the team members' respective skills and learning edges.

So much for personality. How are these extensive development processes related to the maturation of the brain? That is the topic of our next section.

Brain development from 2 to 20 years of age

By the age of 2 years, all three levels of the brain are active and interconnected, but the wiring that connects them is not the same in everyone. All children are born with a given temperament, and the people around the child respond to this temperament as best they can. The child's understanding of himself and the world around him is shaped by these responses, and the ensuing experiences give rise to the countless neural circuits that connect the three major levels of the brain: the autonomic sensory domain, the limbic emotional domain and the prefrontal mentalizing domain. During the second year of life, a comprehensive pruning process takes place in the brain. Rarely used neural circuits die off, while circuits that are in frequent use are strengthened; as a result of this process, every brain develops slightly different skills. The most important condition for further personality development is that the three levels of the triune brain are well interconnected by the age of 2 years, giving the child the basic experience of pleasant and unpleasant sensations and emotions as well as a basic sense of support and security in the relationship with his caregivers.

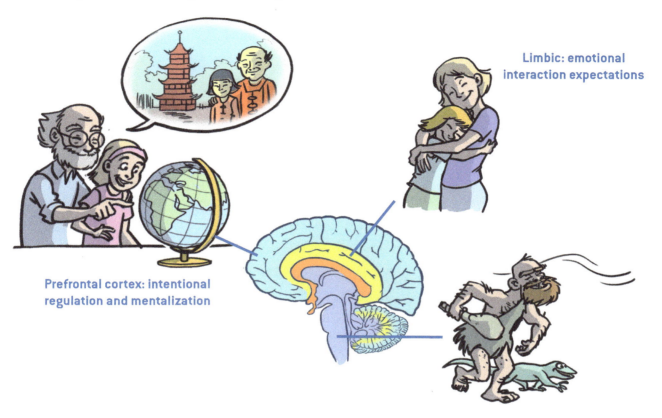

Limbic: emotional interaction expectations

Prefrontal cortex: intentional regulation and mentalization

Autonomic: energy management and body sensations

The extended second maturation process begins around the age of 2 years and continues throughout childhood and youth. All the brain's neural circuits are reorganized and refined in a process that involves the entire brain, beginning with the shorter neural connections and then moving on to the very long ones. These changes gradually enable the child to connect inner sensations and emotions with conscious choice, concrete planning and even abstract thinking.

The maturation process begins in the sensory integration area in the parietal lobe located at the back of the head. It then gradually moves forward through the temporal lobes at the sides of the head, which are crucial for emotional processing. Eventually, maturation reaches the prefrontal cortex behind the forehead, which is central to conscious thought and mentalization. This development continues until around the onset of puberty.

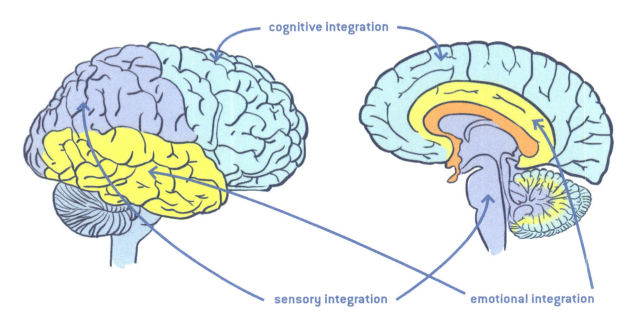

Outer view of the brain   Midline view of the brain

cognitive integration

sensory integration   emotional integration

During the teenage years, this process of refinement continues, accompanied by another pruning period that eliminates unused neural pathways. The brain "trims down" to the level of the adult brain, which has fewer neurons and fewer, but much stronger and more complex circuits. In fact, the brain's development from 2 to 20 years is like a slower version of the extensive growth and pruning that took place from birth to the age of 2 years. Thus, the brain matures in two waves during childhood and adolescence. Each wave begins with extensive growth followed by a pruning period, where only the frequently used neural pathways survive. "Use it or lose it," as the saying goes.

The brain's maturation is largely complete by the age of 20 years, but there is huge individual variation, and different capacities peak at different ages. In fact, the brain continues to mature and change throughout the lifespan. Our thinking and emotions are different when we are 60 than when we are 25 – with progress in some areas and decline in others.

This book mainly offers an overview of a normal or typical development process, although few people actually experience a completely "normal" course of development. But first, we will use the next chapter to take a look at the most common stress factors and natural stress reactions that can occur during childhood, youth and adult life, and which shape our development. In the following chapters, we then return to the normal maturation process.

CHAPTER 2
Stress factors and stress patterns

Stress reactions always occur because we are experiencing far too much input. The function and development of the brain and personality may be overwhelmed by input that is too intense or chaotic, but it may also be impaired and starved by a lack of stimulation. In worst-case scenarios, you may even be simultaneously overwhelmed and understimulated.

Stress factors

From birth to death, we experience three key sources of stress. The best-known source has to do with concrete and practical factors, such as a poor physical environment in kindergarten or at the office. Perhaps the noise level is overwhelming, or budget cuts have led to excessive demands on your skills or working hours. On the other hand, both children and adults may also be understimulated by a lack of suitable tasks or external structure.

Stress can come from having too many tasks or overly demanding ones …

The other two stress factors stem from your social environment and the social motivation systems of attachment and status.

From around the age of 2 years, children begin to be much more involved in social interactions outside the family. These interactions are shaped by the expectations they have developed based on their primary attachment pattern with caregivers. Children with an insecure attachment pattern are likely to recreate that sense of insecurity throughout life. This hampers their ability to build satisfying relationships in the classroom and later in the workplace, in leisure groups or in a marriage. However, most children have more than one experience of attachment patterns, and the experiences of any given relationship will influence their expectations of the next contact. Accordingly, attachment patterns continue to change throughout life. Thus, you bring an attachment expectation with you when you enter into a relationship, but your sense of attachment is also constantly affected by your current interactions. The key to changing attachment insecurity is to engage in the kind of mutual support and challenge that is relevant to the relationship you are in.

... or from a feeling of never really belonging ...

As with attachment, our earliest status experiences come from the relationships within our family and vary in different contexts and across the lifespan. Status issues are a major source of stress for most people, not least because many in our modern culture find it difficult to strike a good balance between democratic equality and relevant authority. Relevant authority takes many forms, such as leadership responsibilities, product oversight or professional expertise, and it comes with power and status. Few people have efficient strategies for handling the difficult emotions that often arise in connection with issues of status. Hence, status exchanges often end up being either too harsh or too soft. Harsh status relations lead to an uncaring, low-trust environment, while too much softness produces frustrating uncertainties about structure and expectations. High and low status are associated with different kinds of stress. Stress in low-status positions often results from the inability to protect oneself from being pushed around, overruled or losing benefits. High-status stress usually involves power struggles, insubordination and unrealistic expectations. Most teenagers and adults find themselves in mid-level status positions where they are exposed to both types of stress.

… or aggressive status behavior.

All three types of stress may be present at the same time.

Three sources of psychological stress or resilience in later childhood and adult life.

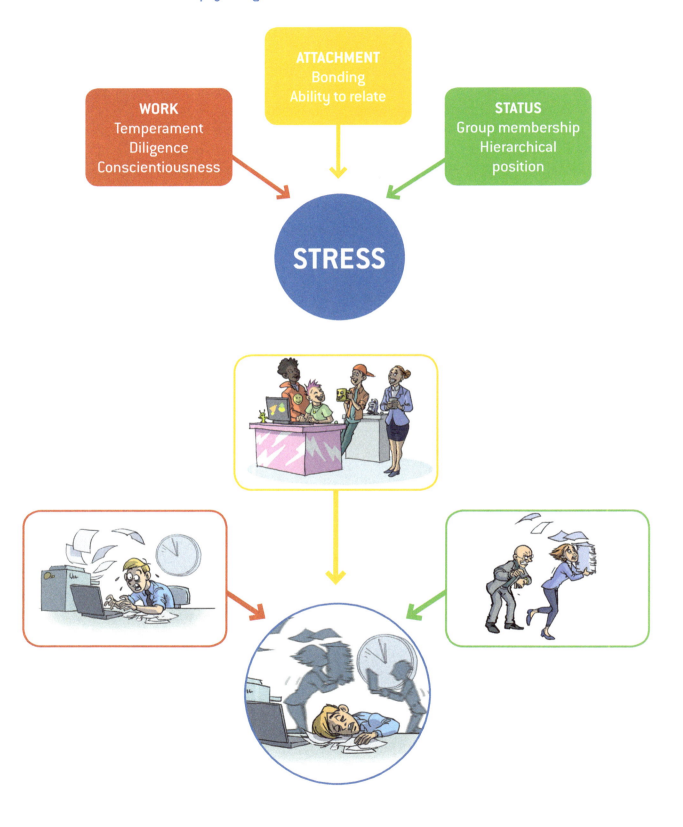

Stress patterns – regression, dissociation and deficient development

Stress and trauma lead to different kinds of stress patterns, depending on when the stressors occur in your lifespan and on their nature, duration and intensity. We distinguish between three different types of stress patterns:

1. *Regression* may be activated at any time of life. Whenever you are temporarily emotionally overwhelmed, you may fall back on the less mature coping strategies you developed early in life.

2. *Dissociation* may be either traumatic or cultural. *Traumatic dissociation* may be triggered when you are exposed to life-threatening experiences. *Cultural dissociation* may be activated when the discrepancy between your basic, innate needs and demands of important relationships or of the prevailing culture is so profound that you are forced to choose between the two.

3. *Deficient development* may look like the other two patterns but is typically the result of severe early childhood stress and trauma; an environment characterized by so much chaos and unpredictability that the child was never able to develop basic emotional competences.

Let us take a closer look at these three patterns.

Regression

Most people have probably experienced situations where the pressure was more than they could bear, and a frustrating incident or demand became the proverbial "last straw". Suddenly, our calm, sensible "normal" self is nowhere to be found, and a more primitive part of our personality takes over. Whether we explode in rage or burst into tears, we are acting out our own individual *stress-induced regression pattern*. Fortunately, the regressive impulses pass when the pressure lifts. We begin to breathe freely again, come to our senses and sort out the situation – until the next flash point occurs. The overwhelming stress of an important exam, a death in the family or being laid off can trigger brief or prolonged regressions or, more generally, give us a "short fuse". However, joyful events, such as getting married or landing a new, important job, can also trigger regressive stress reactions along with the joy and excitement.

Emotional regression occurs when things are just too much to bear …

… but in most cases, we quickly regain control over the situation.

Dissociation

Another common stress reaction is *dissociation*. The cause of *traumatic dissociation* is often sudden and dramatic, such as witnessing a car crash up close, being badly hurt in an accident, being the victim of violence or being subjected to threats of violence. Traumatic dissociation often involves a feeling of mortal danger or of being only seconds from death or serious injury. During the event, you freeze, cannot move, or perhaps you have no time to move. Survival reactions are activated at a level far below the threshold of your normal consciousness, and even though you are likely to go numb, shaky or limp immediately after the event, your organism will usually re-cover from this state within a few hours or days. It helps to have the company of a calm, accepting and empathic person. It also helps to find and re-claim your sense of agency, that is, your ability to choose and to act, in order to feel better or recover your bearings and sense of self. However, in some cases, the impact of a traumatic experience is so deep that it is impossible to return to "your old self". In other cases, you may feel that you got over the trauma without any lasting effects, but inexplicably develop physical or psy-chological symptoms six months or a year later. These post-traumatic states unfortunately have a tendency to get worse over time rather than better – so if you find this happening to you, it is important to get professional help to recover from your traumatic dissociation.

Traumatic dissociation can occur as a result of life-threatening events …

We use the term *cultural dissociation* when the unconscious discrepancy between our natural needs and the social or cultural demands are too great to recognize, much less reconcile. For instance, a woman might live her whole life in a culture where it is considered self-evident that setting boundaries to defend her sense of self would make her a bad person. Consequently, she is always pleasant and accommodating, regardless how unreasonable her situation gets. In early childhood, she lost the ability to appreciate her own anger and sense her personal boundaries. Ultimately, this could lead to depression or to somatic symptoms, such as elevated blood pressure. An unquestioned and unrelenting cultural demand pre-vents her mentalizing prefrontal cortex from noticing and integrating her impulses towards autonomy and boundary-setting and puts her senso-ry-autonomic and emotional-limbic system under pressure, which inhib-its her feelings of joy and may even drive up her blood pressure.

… while cultural dissociation occurs when the unconscious conflict between social demands and our natural needs becomes too great even to recognize.

Deficient or insufficient development

The third type of stress response stems from *deficient development* or, in less severe cases, *insufficient development* and is often overlooked or characterized as a trauma response. The most severe form of deficient development stems from a childhood marred by long periods of chaos, isolation or extreme emotional vagueness, leaving the child unable to develop feelings of attachment and other basic emotional capacities. In the worst-case scenario, he never learns to identify the inner sensations that constitute emotions and feelings. Unable to develop empathy and form emotional attachments, he never experiences joy or satisfaction from intimacy and peer cooperation. Instead, he seeks other sources of satisfaction, for example mindless sex, extreme thrills, status or power.

Deficient development often leads to fundamental distrust of or contempt for the intentions of others. If you do not experience support from relationships, you will often react to high stress with impulsive outbursts or collapse. Children and adults with deficient development are generally unable to acknowledge any responsibility for problems – to them, the problem appears to be that everyone else is unreasonable or crazy.

Fundamental mistrust can easily lead to impulsive outbursts …

Insufficient development is the term we use when the child has had some basic needs met and has developed some basic skills, but not enough to flourish or thrive at later stages of emotional development. These patterns generally develop as a result of inadequate care during the earliest years of life. Perhaps his parents generally had a cold and practical demeanor. Perhaps they provided a relia-

… we can learn to get by without a loving gaze …

ble care structure that met the child's basic needs and developed his basic interaction skills during his early years. They also provided the necessary support for him to develop practical and cognitive skills in later childhood but lacked the emotional capacity to look lovingly at him, comfort him when he was distressed and share his joys and sorrows. He learned to get by with minimal care and emotional support and adapted to very low levels of emotional intensity – just as his parents probably did when they grew up. Childrearing habits are commonly transmitted from one generation to the next.

In another scenario, the child did receive em-pathic care, but only in response to certain kinds of emotions. If his parents only felt comfortable with him when he was happy, he will have trouble recog-nizing or regulating sad, angry or anxious emotions in himself and others. On the other hand, if his par-ents mainly paid attention to him when he was sad or upset, those are the feelings he will be familiar with and identify with in subsequent relationships.

... and if we mainly are used to getting attention when we are sad, that becomes the easiest emotion for us to feel and share with others.

These response patterns are so fundamental that they interfere with the ability to mentalize. Since our basic social abilities are preverbal and acquired before the age of 2 years, a person may have profound shortcomings in these areas and still be highly intelligent and articulate.

Insufficient development may also stem from experiences in later child-hood and adolescence. For instance, if your family moved often, you may never have had a chance to develop deeper close friendships with peers. Or perhaps your childhood and adolescence unfolded under the shadow of se-rious illness in your family. Maybe you were naturally timid and never got sufficient adult support to learn to step up or try new things. Or maybe your parents were so focused on your needs and happiness that you never learned to obey rules and be considerate of others. Perhaps you had health issues that prevented you from learning to co-create games and game rules with peers.

This brings us to an important point: Developmental failure and trau-matic dissociation can look quite similar. Both can lead to emotional de-tachment and hypersensitivity, and while there is a huge amount of literature about stress and trauma, deficient development is rarely described. Hence, it is easy to assume that there are "frozen" personality resources just waiting to be uncovered, which will unfold naturally if only the person works through the trauma. The problem with deficient development is that there are no hid-den or frozen resources. What you see is what you get. The only way forward is to begin to develop the resources and skills that the person lacked the sup-port to develop in the first place – and that takes time.

Let us for a moment compare personality to a sweater, as a metaphor for the three stress responses. *Regression* is like having the sweater pulled out of shape – it needs to be pulled back into shape after the next wash. With *traumatic dissociation* there is a big tear in the sweater, and we need to pick up some stitches and repair it. With *deficient development* we have to start from scratch, spinning the wool to make yarn, and with *insufficient development* we may have just half a sweater to work on.

regression   **dissociation**   **deficient or insufficient development**

Finally, it is important to remember that some people struggle with all the stress factors and stress reactions at once – deficient or insufficient development, trauma, regressive responses, culture shock, lack of relevant competencies and language and insufficient attachment and status. This is a tragically common state of affairs for refugees or others who have been transplanted to a very different culture. They may be unable to speak the local language and have none of the skills that matter in the new community. Often they have no friends or extended family. At work – if they manage to find work – they are at the bottom of the pecking order, which is also where they tend to end up in social settings. Apart from these immediate challenges, they may be burdened by recent trauma as well as an emotionally impoverished childhood. This combination is fairly common for some of the refugees fleeing from war zones and famine, who are trying to build a new life in the West.

We may struggle with deficient development, cultural dissociation and traumatic dissociation …

With complex stress and trauma, the best approach to therapy or social integration will vary from person to person. Studies the world over have found that two of the most helpful elements in helping people to recover after disasters and trauma are attachment and agency: belonging to a social community and having enough support to act to improve one's situation. Both depend on supportive relationships with other people, so in the next chapter we take a look at two social learning contexts: *symmetrical* and *asymmetrical* relationships.

… and still have to find a way to build a new life, without the basic skills of belonging and usefulness that are taken for granted in our new country.

CHAPTER 3
Symmetrical and asymmetrical relationships

For a healthy development we need both *symmetrical* and *asymmetrical relationships*. In symmetrical relationships, each person is equally responsible for the relationship and thus has to be fairly self-reliant. That is the case for social learning among children and young people, for example. The capacity for handling exchanges with peers emerges around the age of 2 years, when children – initially with adult guidance and supervision – begin to play with peers and need to work out how to share toys and candy.

Sharing is hard when you are 2 or 4 years old – and sometimes even when you are a grown-up.

As we mature, the symmetrical responsibility grows. As teenagers, we have a greater responsibility for looking after ourselves and acting sensibly than we do when we are younger. Young people have to be able to meet people around them with relevant support and reasonable demands, and they have to be able to help build well-functioning groups. Over time, symmetrical relationships become increasingly important, as we develop attachments with classmates, teammates, friends, boyfriends/girlfriends and, as adults, with coworkers and our own parents and adult children.

In an asymmetrical relationship, one person has the primary responsibility. For instance, a caregiver is responsible for a child. Parents and preschool teachers are responsible for establishing a generational hierarchy and creating a nurturing environment for the development and well-being of the children in their care. Asymmetrical relationships are not reserved for childhood – throughout your teenage and adult years, you can develop deep insights from exchanges with an insightful mentor or teacher, a coach or a psychotherapist. You can seek out loving and insightful individuals who exemplify ways that your life could hold more of what you yearn for: joy, love, open-mindedness, clear boundaries or other qualities. As you mature, you too may act as a mentor or teacher for someone else, thus gaining experience with the position of authority and responsibility in an asymmetrical relationship.

During difficult times, you can get help from loving and insightful people and even internalize them as implicit role models.

Self-agency – your personal mastery in positions of authority

You cannot teach what you have not mastered. If you are responsible for another person's learning, you have to master the skill in question at a higher level than they do. In emotional development, the term *self-agency* refers precisely to this ability to use yourself and your own skills to help someone else learn. To teach someone to dance, you have to know how to dance. You also have to be able to dance in a way that makes it fun and exciting for your student to learn. Forget "Do as I say, not as I do" – simply describing the steps will not be enough. The same is true of social skills. It is not enough to be able to explain them. It is not even enough to be able to demonstrate them. We have to be able to co-create them with the student in engaging interactions in order to create the harmony and rhythm of mutual contact.

What you think you are doing is not always what you are actually doing ...

Most of us have no idea what sort of social skills we possess, nor do we remember how we acquired them. This is because we learn the most basic interaction skills with our parents and playmates before we are 3 to 4 years old. Since most people have a rather fuzzy sense of what those skills even are, it is difficult to assess our own strengths and weaknesses. It is equally difficult to spot the strengths and weaknesses in others and to determine their learning level.

Your nonverbal consciousness stores your actual interactions with others, while your verbal consciousness describes them in words. When you are lucky, those words match your nonverbal experience, but often they do not. That makes it important to have the courage to explore your interaction skills, approaching this process like a game or a journey of discovery. This could include making short videos of yourself interacting with others, both in situations where you have a position of responsibility and in equal, symmetrical relationships. The best way to assess the interactions is probably to enlist the help of a professional, so you can discuss what you observe. The development chapters later in the book list some of the skills you can explore in this manner.

... and becoming aware of the difference can be an uncomfortable experience.

The learning zone – the zone of proximal development

When we speak of social skills – our own, a child's or another adult's – there is one recurring question: How do we determine the right place to begin – where is the "zone of proximal development?"

In fact, the development of social skills is driven largely by the same principles as the development of language skills. For example:

> Like language, social motivation systems develop through interactions with others.

> Like language, everyone uses motivation systems differently and has more resources in some areas and fewer in others.

> As with language, children first learn to master simple exchanges and later move on to more complex ones.

> Both in language and with regard to social motivations, the simple skills remain in use for the rest of your life and form the building blocks for more complex abilities.

For instance, you begin to practice turn-taking during infancy. You do so by alternating between expressing yourself to someone else and quietly paying attention in order to capture all the details of what the other person is coming up with, while your body and face spontaneously make lots of little affirmative movements. You continue to do this throughout life, generally without giving it much thought – except when the communication goes haywire. That can happen simply because your social skills are so different from the other person's that you misread each other. The other may be accustomed to expressing agreement by beginning to speak before you have finished your sentence; however, that may make you feel that you are being cut off. A given skill, such as using eye contact to check what the other person thinks of a situation, may also be completely automatic in one person and completely unfamiliar to the other. The way you speak – your tone of voice, facial expressions and body language – has a huge impact on the way others experience interactions with you.

Others' perceptions of you are shaped more by the way you express yourself than by your words.

Development does not end with childhood. All the developmental stages you go through in childhood and youth exist in an adult version, too. Just as you never outgrow turn-taking, you do not outgrow your other basic skills. If you failed to learn a particular skill, you can still begin to develop it later. Interpersonal skills are like language skills; you can improve them at any time of life.

However, to learn something, regardless of the skill, we have to start at the level we are at. That level is called the *zone of proximal development* or the *learning zone*. The learning zone is just one step away from what we already master. This may seem obvious, but often we focus on the things we are unable to do, as in "You have to learn how to [enjoy small-talk, be a good listener, and so on]!" Unfortunately, that will not tell us what our learning level is. Regardless of biological age, a person can only develop emotional competencies at the level they have reached, and it is up to us as mentors to find ways to make the activities of the learning zone meaningful. The person's ideal goal for social skills may well be a long way from his current skill level. On the other hand, as mentors we must determine what level the learner is at in order to find the learning zone. His current mastery determines the next possible step. Social games and activities that match a person's learning zone feel engaging and exciting to him – he has to stretch to master them and experiences new victories, big and small, as he improves his ability to relate to others.

Four dimensions of interactions within the learning zone

The ideal setting for learning social skills involves interactions with a balanced mixture of *structure*, *engagement*, *nurture* and *challenge*. In symmetrical relationships, such as those between children or young people at the same level of maturity, balancing the four dimensions is a shared responsibility. In asymmetrical relationships, the mentor or teacher is responsible for establishing the following aspects of the interaction:

> *Structure*: The teacher establishes good and consistent routines and satisfying ways of being together. The framework is familiar and predictable. The teacher creates opportunities for synchronization and turn-taking in the interaction.

> *Engagement*: The teacher feels and expresses enjoyment and enthusiasm in the interaction and creates interactions and activities where the learner feels the same way.

> *Nurture*: The teacher shows care for the learner, whether in a friendly chat over a cup of tea or by expressing concern if the learner is unhappy or upset. At many stages of development, the learner may also need to learn to be more caring towards others, so the teacher may accept relevant nurture from the learner.

> *Challenge*: The teacher offers appropriate challenges and tweaks them to make sure they remain within the learner's learning zone.

To assess a learner's mastery of social skills, the teacher needs an understanding of individual areas of social skills and their developmental steps. To initiate that process, we devote the next chapter to the five areas of social skills: the social motivation systems.

CHAPTER 4
The social motivation systems

In the introduction we briefly outlined five social motivation systems:

> Attachment
> Play and cooperation
> Hierarchy and status
> Gender identity
> Mentalization (the capacity for empathic insight into oneself and others)

In this chapter we describe the systems in more detail as a basis for identifying the developmental stages.

Attachment

Attachment provides the sense of security you feel in your closest relationships throughout life. Primary attachment patterns are formed during the first year of life, and every human being is born with the need to form attachments with caring adults. Being securely attached to a parent gives the child a secure base, which in turn enables him to venture into the world. During his adventures, he will repeatedly return to mom or dad for a quick moment of emotional reassurance. As soon as he is ready, the parent helps launch him into his adventures again. This attachment process continues throughout life, from secure relationship to challenge and back, with parents, siblings, friends, partners, coworkers and eventually your own children.

If a child feels uncertain about an activity, he may also seek reassurance through eye contact, in order to check out how his parent – or friend – feels about the situation. This is called *social referencing*, and like other early contact skills it stays with you for life. When something happens in your surroundings that startles you or makes you uncomfortable, you seek eye contact, ideally with someone you trust – but even a stranger might do in a pinch. Social referencing also builds group cohesion. You seek eye contact with people you feel connected to – and shared eye contact increases feelings of connection.

Another aspect of secure attachment is the ability to *re-attune*, that is, to reestablish closeness after a disturbance or a conflict, a so-called *misattunement*. Experiencing that misattunements can be resolved and repaired increases your sense of security. In fact, it is one of the two core building blocks of basic trust in life. The other is *interdependence*; the ability to be comfortable with both intimacy and independence.

Insecure attachment patterns are the result when something goes wrong in our most important early relationships: a child's relationship with his parents. Parents cannot give the child a feeling of security that they themselves never had. Intense stressors, such as long-term health problems, can cause profound damage to the attachment bonds within a family – sometimes for generations. Social exclusion and bullying in a family or peer group can also lead to a profound and lasting wound to one's sense of belonging.

If your own primary attachment pattern is secure, you will probably still recognize the feeling of insecure attachment from your own occasional experiences with insecure interactions. It is only when most of your interactions are insecure that they shape your close relationships and your sense of being-in-the-world. Security breeds security, just as insecurity breeds insecurity.

Children and adolescents tend to base their interactions on the attachment pattern they developed in their family. However, they may develop new attachment experiences with playmates, friends, romantic partners and peer groups. Between the age of 6 and 9 years, these different attachment patterns fuse into a primary pattern, which represents the child's best current capacity for dealing with others. In stressful situations, however, less functional attachment patterns will often emerge. Let us take a quick look at secure and insecure attachment patterns during the teenage years.

When something startles or worries you, you try to catch the eye of someone you feel connected to.

With a *secure attachment pattern*, you feel safe in a close relationship — and you are also comfortable doing things on your own. You are realistic and confident about your ability to resolve conflicts with partners and friends.

With an *insecure avoidant attachment pattern*, you are most comfortable in a relationship with a safe distance and plenty of room for differences. Intimacy quickly comes to feel stifling or overwhelming, and you may respond to conflicts by seeking more distance.

With an *insecure dependent attachment pattern*, you are most comfortable in a relation-ship where the two of you are together most of the time and share everything. Differences and distance feel threatening, and you may respond to con-flicts with an intense desire for closeness.

With an *insecure ambivalent attachment pattern,* you are most comfortable in a relationship that alternates between intense intimacy and intense conflict. Lack of intensity feels threatening, and you may respond to conflicts by wanting to dive into them — but find them difficult to resolve.

With an *insecure disorganized attachment pattern,* you are only comfortable in a relationship where you have total control. Loss of control feels like chaos, or even like a threat to your life, and you may react instinctively with sudden violence, flight impulses, numbness or terrified collapse.

Play and cooperation

All mammals – including humans – acquire vital social skills through play. Play is essential during childhood and youth, but it is important for adults, too. When we play, the neurotransmitter dopamine activates joy circuits in the brain. Everything we learn through play is stored in strong memory tracks, and scientists have found that being absorbed in play is a great way to reduce anxiety and depression and boost joy, attachment and socially adaptive behavior. Play is also a driving force during work.

By playing with his parents, a baby develops the basic building blocks for playing with other children later. Early peer play is physical and consists mainly of imitation and turn-taking, such as peekaboo-type games, running and jumping together, playing tag or rough-and-tumble. Gradually, children begin to imitate and share activities they have observed in adults and older children. This develops into role-playing, as the children take on different roles in imaginative play. Later yet, play takes the form of structured games with common ground rules, such as ball games and song games.

Through a mix of turn-taking, imitation and creative contributions, children create a shared intense joy-filled energy and process. However, there are many stones in the road! Maybe the toy is too exciting to share. Maybe one child offers an idea that the other disagrees with. Maybe one of the children does not follow the rules. Maybe something happens that feels profoundly unfair. In these cases, shared joy collapses into frustration. The same process occurs in adult life. Misattunements are just as intensely frustrating as the game is exciting. Hopefully, grown-ups will be able to sort themselves out, but children need adult supervision. Sometimes they can reattune on their own, sometimes they need help. When that happens, a kind and firm authority figure has to help them find a way to get back on track.

Misattunements are incredibly frustrating, and children often need a friendly authority figure to step in and help.

Throughout life, play impulses make you take on new and exciting challenges. Through play you learn to cooperate, and the desire to play gradually develops into creativity, both in your work life and in your personal life. You have probably experienced the deep frustration that both creative processes and cooperation can activate. The wonderful experience of getting back into the flow – alone or with others – can help you when the next meltdown occurs. Having access to the vibrant energy of play is key to increasing your reattunement skills and your tolerance for frustration.

Hierarchy and status

All social mammals establish group hierarchies. That is useful, because it defines ground rules where everyone knows what their status is, who has the right of way, who can be pushed around and who is currently competing with whom. In evolutionary psychology, status is defined as having priority access to resources. Among children and adolescents, this means that if no adult claims leadership and takes charge, the status relationships in the peer group determine who runs the show, gets the best goodies, and has the authority to exclude lower-status kids from the group.

Your inner experience of status is reflected in your body language. Although different cultures have slightly different ways of indicating high and low status, people generally express high status by puffing themselves up and displaying confident movements and facial expressions, while low status is expressed by making oneself smaller and displaying apologetic gestures and facial expressions.

You express high status by making yourself appear bigger and using confident facial expressions and a confident body language, and low status by making yourself smaller and showing self-effacing facial expressions and gestures. This happens unconsciously; doing it deliberately is less effective.

Status is often described by using concepts such as respect and dignity. You may "command respect" or "feel respected". You may feel that you or others "are dignified" or, embarrassingly, have "lost face". High status can be expressed either through threats and aggression or through kindness, caring and generosity. The child gets his earliest experiences of hierarchy and status in his family. Ideally, his parents will be comfortable with their superior status in the family hierarchy and use it to provide a clear, nurturing and sensitive framework for him. That will give him a healthy "inner map" for navigating status relationships with adults and peers in the world outside his family.

Your inner sense of high or low status is not determined by objective factors but by your subjective experiences. The experience of high status is associated with a lower stress level and higher levels of *serotonin*, a neurotransmitter that produces a sense of well-being and calm. Physiologically aggressive or dominant behavior in both sexes is also associated with a higher level of male sex hormone, *testosterone*. So is higher status. When you win a competition that improves your status, your testosterone level goes up, which may then trigger competitive and dominant behavior.

The experience of low status is associated with higher stress levels and with the stress hormone *cortisol*. Physiologically, low status will make you more reactive to stressful situations and it will take longer for you to calm down afterwards. You may also have a strong impulse to attack lower-status individuals, since that tends to lower physiological stress levels.

Generally, boys have been considered more likely to use physical aggression to achieve dominance, while girls have been seen as more likely to use social means of power, such as gossip or bullying. Recent research paints a slightly different picture. Boys commonly use social power in status relationships, while it holds true that girls rarely use physical force.

Gender identity

Gender identity can be broken down into three aspects, which develop at different times. The first is *core gender identity*, the physical experience of being a boy or a girl, which begins to form prenatally and is set at the age of 14 months. Studies show that a disposition for same-sex attraction may be generated during this phase, although it can also form as a result of later experiences and cultural influences. The second aspect is *personal gender roles*. The primary window of development is from around 2 years to 10–12 years, although gender roles remain open to change throughout adolescence and adult life. *Sexual maturity* is the third aspect and occurs during puberty.

In general, boys develop slightly later than girls. The prefrontal cortex is not fully mature until we are in our twenties, and typically it matures a little later in men than in women, although there are significant individual differences. Brain and personality developments are also affected slightly differently by male and female sex hormones.

Male sex hormones intensify the effect of two neurotransmitters that boost energy and combativeness: *dopamine* and *adrenalin*. For this reason, boys often need a great deal of physical activity to be well regulated, and their play is more likely to revolve around competition between individuals or teams. Boys are also more accepting of physical aggression between peers. Girls, on the other hand, are generally better at polite pretense.

It is usually harder for boys than for girls to pretend they are excited about a boring present.

The female sex hormones boost *serotonin* and *oxytocin*, which have a calming effect and enhance bonding and quiet, pleasant interactions between close friends. Oxytocin also increases the experience of "us versus them" and hostility towards "them". When girls play with other girls they talk to each other about what they are doing and pay attention to how the others are coming along. Girls of all ages sit closer together and make direct eye contact with each other a lot of the time, while boys sit facing the same way and rarely look directly at each other. However, gender differences are most evident when the opposite sex is not around. In mixed groups, individual differences are more noticeable than gender differences. Scientific studies show that gender-specific behavior only accounts for about 10 percent of differences in behavior.

Mentalization

As far as we know, we humans are the only living beings capable of shaping stories, *narratives*, about our lives and experiences. Already in their second year children begin to use symbols and language in play and to communicate their experiences. With both children and adults, narratives may be more or less coherent and more or less factual. It is important to realize and remember that all narratives reflect our subjective experiences rather than objective reality.

Mentalization is a technical term for the process of understanding and empathizing with our own and others' impulses, emotions, thoughts, experiences and actions. Emotional intelligence and mindsight are popular terms for this capacity, and the only way to get better at it is by interacting with others. It is not easy. The main challenge in mentalizing your own experiences is to step outside your own perspective in order to "see yourself from outside" – not in a judgmental way but with interest, understanding and kindness. When mentalizing others, the main challenge is to really put yourself in the other person's shoes – "seeing the other from inside" – with interest, understanding and kindness.

Well-developed mentalization is characterized by curiosity, humor and reflection. You are open to the unexpected and unknown while maintaining a healthy skepticism. You are empathic, and you have a pretty accurate sense of how the other person perceives a given situation. You are aware of your own inner reactions and also notice how the things you say and do affect others. You are also aware of and curious about mental processes: how emotions generate behavior, how thoughts influence actions and whether a behavior or feeling seems unconscious – in yourself as well as others.

Unfortunately, nobody operates at this level of sophistication all the time … We all have features from earlier levels of maturation. Maybe it is difficult for you to recognize emotions. Maybe you get emotions, sensations and thoughts mixed up. Or maybe you only really notice and react to concrete actions, and do not pay attention to other people's feelings and intentions. Maybe you are totally convinced that you know what others really feel, no matter what they say. You may not have had the opportunity to develop your mentalizing skills fully. Having underdeveloped mentalizing skills makes it difficult to reality check your sense of a situation and to understand how your actions affect others.

Noticing only concrete actions, believing that others know and think the same as you and confusing fantasy and reality are natural childhood stages of mentalization. In adults, stress is likely to bring out one or more of these early patterns. No one is born with good mentalization skills. It is the last of the human motivation systems to develop, and it can continue to grow and deepen throughout life.

Normal mentalization skills typically develop from a secure attachment pattern. However, exceptional mentalization typically grows out of much more painful experiences, involving both secure attachment and intense challenges that the child was only barely able to manage. It appears that what matures us most is hardship – provided it lies within our learning zone.

Seeing others from inside … and ourselves from outside … with kindness.

The nexus of attachment and mentalization forms a central component of our personality. A secure attachment pattern enables us to sense ourselves and seek support in a crisis. Good mentalization enables us to see difficult situations in a broader context and come to terms with them. With secure attachment, it gives us the personal strength and flexibility to handle painful experiences.

This completes our overview of the social motivation systems. In the next three chapters we outline their gradual development. Each chapter describes one of three main periods of socialization: preschool (2–6 years), early school age (6–12 years) and adolescence (12–20 years). It is important to note that each developmental level in the following chapters is like a step on a set of stairs. The goal is not to race to the top and stay there but to be able to use all the steps. You rely on your interaction childhood skills every day of your life.

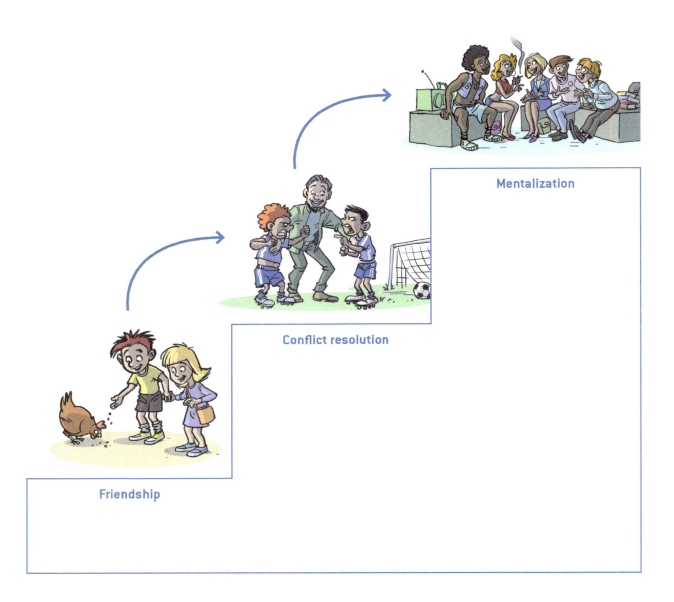

Mentalization

Conflict resolution

Friendship

CHAPTER 5
Fantasy, roles and social codes

Personality development from 2 to 6 years of age

Between the ages of 2 and 6 years, a major shift occurs in the child's social skills. Before the age of 2, the child mainly relates to adults and learns social behaviors from them. From this age, however, interactions with other children rapidly take on much greater importance. Also, the child is now beginning to connect his body-based experience of the world with cognitive symbols, in the form of pretend play, words, drawings and role-playing.

Once the child is able to understand and use symbols, an object or a word can evoke his experience of the thing that it symbolizes; he has an *inner working model*. This means that a toy dog or the word "dog" can activate the child's experiences with the real animal. The ability to engage fully with symbols and roles is crucial because it links your experience of reality to your verbal and cognitive map. It is this link that gives life and emotional meaning to pretend play and words.

There are many levels of understanding in symbols and pretend. At the beginning of Antoine de Saint-Expéruy's classic book *The Little Prince*, the narrator, a downed pilot, describes his childhood habit of testing grown-ups by drawing them a particular picture and asking them what they saw. Most replied that they saw a hat. They understood that he had drawn a symbol but did not have the imagination to go deeper. The little prince, however, immediately recognized that it was a python that had swallowed an elephant. At last, the pilot had found someone who shared his way of seeing the world.

What do we see – and feel – in a symbol?
Is this a hat …

… or an elephant inside a python?

Symbols and inner working models can have many dimensions, and for both adults and children, having an inner experience of them is central to connecting with other people. Without this skill, you have no way of thinking about the feelings of others. Without it, you would not be able to soothe yourself with elaborate daydreams – or fantasies of the perfect retort. You would not be able to imagine alternative ways out of an impasse or come up with a creative way to resolve a conflict. You also could not develop the appropriate roles and expectations for differing contexts; without these inner models, you would have no way to distinguish between the different behaviors that are appropriate at home, in school and in after-school settings. Examples from adult life include appropriate behavior at work, your interactions with your own children and the way you relate to authority figures, such as a doctor or a police officer.

We use language differently in different roles. Linguists have found that bilingual people change their behavior and self-perception when they switch languages. This is known as *code-switching*, and we use the same ability when we modify our behavior to fit into different social settings. Code-switching requires good inner working models as well as a command of the different interaction "languages" that are suitable in different contexts. Children and adults who have not mastered code-switching are socially challenged.

Brain development from 2 to 6 years of age

By the age of 2 years, the child's primary sensory and motor regions have reached their adult size, and the primary neural pathways between the brain regions have been established. A new maturation wave now begins at the back of the cortex, in the visual processing center within the occipital lobe and in the sensory association areas within the parietal lobe. One key development during this process is the formation of fatty myelin sheaths that speed up neural signaling by insulating the nerve fibers; another is the growth of vast numbers of synaptic connections that refine neural communication.

The *insular cortex* in the brain has been called the seat of the "mind-body connection". It is active when sensations, emotions and thoughts feel important and emotionally meaningful to us, for example when you feel "a lump in your throat" or have "butterflies in your stomach". The insula is also connected with Wernicke's and Broca's areas, the seat of language comprehension and production, which blossom with the child's rapid development of language and symbols. This insular-language-area connection improves the child's ability to understand and use language and to read the emotional expressions in the voices of others. There is also a substantial expansion of the connections between the frontal lobes, which handle internal representations, symbols, thinking and impulse control, and the parietal lobes, which handle our perceptions of ourselves and our surroundings. The connections between this frontoparietal system and the insula also expand significantly. The insula is central to the feeling of disgust, which can now be evoked by moral thinking, triggered by acts that we see as bad.

The insula – the mind-body connection

From 2 to 4 years of age

When a child plays with an adult, the play is asymmetrical. The adult protects the child and adjusts to his skill level by making the activity less demanding. Like all mammals, human parents self-impose "handicaps" in order to let the child come out on top and feel like the stronger and more competent of the two. Around the age of 2 years, however, the child begins to establish himself in a new world of peer relations, and the adult helps him along by taking more of a back-seat role. Ideally, the child now has his first symmetrical attachment experiences in exciting games with his favorite playmates. With these new friends, the shared engagement in pretend play or a joint project provides deep satisfaction. Friendships that are established during this period can last for years.

The importance of play for social development can hardly be over-stated. The earliest forms of peer play are physical and often involve imitation. Children may roll a ball to each other, jump on a trampoline together or play side by side in the sandbox. Imitation and repetition are vital to developing shared emotions and a shared focus, as children who play together learn to hold the same play scenario in mind. Imitation alone, however, soon gets boring. Around the age of 3 or 4 years, improvising and building on each other's initiatives become a central part of play. If one child builds a small tower out of toy blocks, another child may pick up the theme and build a bigger tower, while a third child might add structures that tie the towers together to create a small town.

In quality play and cooperation, you have to be able to come up with an idea …

… build on someone else's contribution …

… and bring everything together to form a larger whole.

Creative development occurs in all play. It is at the core of the human ability to cooperate on big projects. It only works, however, if the feeling of contributing to a joint venture is strong enough. If individual initiatives outweigh teamwork, the project suffers and may be abandoned. On the other hand, if there is not enough individual initiative, the shared activity will tend to get boring and, again, be abandoned.

It is in these shared games that children develop early forms of mentalization. The better they are at working along with each other's intentions and ideas, the more fun they can have. In song games, dance moves and gestures are associated with the story line in the song, and children begin to be able to imagine the different roles or events in the song. Around the age of 3 years, children begin to develop elaborate forms of pretend play. They fashion a shared fantasy world and narrative, such as playing circus or building a farm out of toy bricks. By grasping others' ideas and cooperating they can sort out what does or does not belong in the shared pretend world. What happens if you introduce a helicopter into a farm narrative?

Can a helicopter fit in on a farm?

Peer play is symmetrical, and unlike grown-ups, peers don't self-handicap to make each other feel better. This means that children must learn to:

> maintain joint attention on a project
> share and share alike
> take turns
> follow others' initiatives – and offer their own
> manage their frustration with snags in the process
> be able to reconcile and reattune after misattunements

Children who are not skilled enough at these aspects of play risk being excluded or pushed around. Symmetrical play is a ruthless setting for socialization and training in the collaborative skills you need for the rest of your life. For adults, too, it can be tough to give up your own perspective and be open to others' ideas, emotions and perspectives. It is cumbersome to have to explain things to others in a way that they understand, and which inspires them to "play along". Fortunately, the rewards generally outweigh the effort. A vast new world of joint activity opens up as you get better at communicating and developing shared visions and mental worlds.

Sometimes, two pretend worlds will clash …

Sometimes, children are overwhelmed with frustration. If their language and emotions are sufficiently matured and interwoven, they may be able to draw the line by shouting "No!" rather than hitting or biting. They also quickly learn when to call in a grown-up to help them establish a fair structure and get back on track.

… unless there is someone at hand who can help them reconcile.

This kind of support and care from adults outside the family gives the child confidence that help is available in the bigger world. It also helps him develop methods for managing frustration and resolving conflicts with peers.

During these years children also begin to develop their sense of right and wrong. Four-year-olds can even distinguish between different *kinds* of right and wrong. They are not yet able to explain this difference in words – it is a gut feeling – but they understand that while it is not quite right to introduce a helicopter into a farm theme, it is much worse to take a toy away from another child.

You can do something wrong by accident … or on purpose.

At this stage, the child's mentalization capacity is sufficiently developed to consider intentions. This means that it makes a difference whether someone did something wrong on purpose, or whether it happened by accident. With adult help, children are now able to establish common ground rules, such as the rule that everyone has to ask nicely for a toy instead of simply grabbing it.

However, no one does the right thing all the time, even if they know what the right thing is. Often, you will experience a conflict between what you feel like doing and what you need to do to get along with others. At this age level, children develop important new emotional skills: a conscience and an awareness of guilt. These capacities can only emerge if the child has developed an inner sense of right and wrong. They stop the child from simply following his immediate impulses and compel him to do the right thing – when they are strong enough. Meanwhile, he also develops the ability to deny any wrongdoing and to conceal inappropriate emotions in order to avoid punishment, shame or guilt. That skill in turn expands the child's mentalization capacity, as he begins to understand that others, too, may lie, make mistakes or do bad things.

Once you have managed a difficult situation by telling a lie, it gets easier to spot when others lie.

From 4 to 6 years of age

At 4–6 years of age, most children have found regular playmates that they get along with particularly well. They have favorite games and have had innumerable emotionally charged experiences, including both smooth and pleasant play and countless conflicts. Their sense of time is developing as well, and with that comes an emergent sense of cause and effect.

They can now follow a storyline and create longer, coherent narratives, both in pretend play and when they talk about real-life experiences.

Individual friendship attachments are now in place, and the child is developing a sense of group belonging, which expands his attachment pattern to include the groups he is a member of.

While both boys and girls can enjoy ball games and playing on climbing frames, stereotypical gender roles are also prominent in pretend play and role-playing. Boys might use toy blocks to build police stations or play some version of cops and robbers. If they dress up, they are monsters, zombies or superheroes. Girls typically engage in social play, often with a focus on caregiving or looking alike, and if they dress up, they are more likely to dress up as princesses or witches. It is important to realize that there is nothing wrong with stereotypical gender roles. Children pick up typical patterns from their family and from the world around them, such as films, TV programs, video games and social media. In pretend play they rehearse stereotypical interactions and expressions, including gender stereotypes. The same is true of body language, such as specific ways of walking, standing and sitting. These stereotypes help you learn the behaviors generally associated with certain roles and situations. A sense of stereotypical patterns is an important element in mastering code-switching.

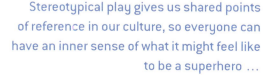

Stereotypical play gives us shared points of reference in our culture, so everyone can have an inner sense of what it might feel like to be a superhero …

… or a mom.

In all-girl or all-boy groups, hierarchy and status are expressed in different ways. Boy groups usually have a clear leader who issues orders, and boys often have several minor confrontations and spontaneous competitions. For them, a clear goal for the play project and explicit leadership are key priorities. Girl groups typically try to avoid conflicts. They usually make decisions by offering suggestions or making requests and use negotiation to reach consensus. Feelings of caring and closeness are the top priorities.

Status in a group depends partly on innate qualities. Even among 3-year-olds, some children tend to lead, while others mostly follow. Studies have found that already at this age, leaders have a good deal of power when the children choose play activities. They can permit or prevent other children from joining a game. Innate sensitivity or a willingness to take risks combined with a high energy level may lead to high status in some groups, but low status in others. Acquiring relevant social skills is also important for status: Does the child know how to interact with other children and play in a way that matches the group's norms?

Language, peer attachment and a sense of time all expand the child's ability to mentalize. Children can now remember past problems and solutions and anticipate possible future difficulties. Accordingly, they will begin to sort out rules and guidelines before they begin to play or call a time-out in a game in order to sort out ambiguities as they emerge – when they remember to do so. With these skills they can begin to suggest solutions to conflicts and sometimes even find ways to reconcile after arguments without assistance from grown-ups.

It is not always easy to agree on the storyline, but children gradually learn to resolve conflicts without adult assistance.

Children are now beginning to imagine how others feel and empathizing with their situation. That enables them to understand the emotional processes at play in other children's interactions, as in, "Mary was upset because Nina took her doll, but Nina didn't know it was hers." This kind of reflection develops moral thinking. Morals are not only about rigid rules – they grow out of the experience of the connection between emotions, intentions, causes and effects.

Children also begin to be able "to see themselves from outside". They may engage in simple reflections on how others perceive them, and they begin to describe themselves in material terms, as in, "My name is Peter – I am almost 6 years old, and I have a red bike." With this growing sense of reality, they also begin to realize that they do not excel at everything ...

Sadly, you are not Superman or the world's best soccer player.

Exploring fantasy, roles and social codes in yourself and others

The social and personal skills that develop at the age of 2–4 and 4–6 years can be more or less developed in older children or adults. If you would like to see whether this is a zone of proximal development for yourself or someone else, you may use the questions below for inspiration. They are phrased as invitations to self-reflection because it is important first to examine those areas in yourself, even if your focus is on supporting the development of others. Assessing your own personal and social skills requires more than your own opinion of them, and assessing them in others involves more than simply asking questions. For all of us, there is a gap between what we think we do, and what we actually do. In order to dig a little deeper, you may either examine a specific situation, ask friends who know you well or take a few days or weeks to gather information on a particular question.

Attachment
> Do you have favorite coworkers (playmates), that you trust? Are you comfortable working both independently and in close collaboration; and are you able to resolve the conflicts that emerge?

Play and cooperation
> Are you able to collaborate smoothly on a task (or game), either in pairs or in a bigger group? Do you tend to take charge, or are you more likely to assist or follow?

> Are you able to play with symbols, flights of fancy and ideas with others, or do you lean more towards the factual and rational?

> Are you able to engage in a joint process to co-create a coherent description of an experience or a situation?

Status
> When you are with others, are you often assigned the leadership position, or does it usually go to others?

> What position do you prefer? Are you happy with the position you usually find yourself in, or would you prefer a different one?

> Do you think there is a difference between the position you have in different contexts, for example at home and at work?

Gender identity
> We all have both feminine and masculine aspects. Find some examples of your masculine aspects – and then some of your feminine aspects.

> Do you think your interests are mostly masculine or feminine?

> Which of the two do you think is your dominant aspect?

Mentalization
> Try to make three factual statements about yourself.

> Now make three statements about your inner character, what you are like as a person.

> Are you able to "see others from inside"? Do you notice how others feel and take the time to learn what they enjoy and dislike, and what they find easy and difficult?

Connection between body and language
On this subject, it is not enough to consider the questions, you have to experiment a little.

> Take a moment to recall the questions you just reflected on. What emotions and sensations did you notice?

> Close your eyes for a moment, and sense one of your hands.

> When you have a clear sense of the hand, then open your eyes and look at it. Does the sensation in the hand change? In what way? Now say, "my hand", and notice how the words connect – or do not connect – with your sensation of your hand and with your experience of looking at it.

> Repeat the process with your other hand. Is there a difference in your experience of the two hands?

CHAPTER 6
Group norms and realism

Personality development from 6 to 12 years of age

The key element of emotional development in 6-12 years-olds is the expansion of mental perspectives and horizons. The child gets much better at delaying gratification, regulating emotions and exercising volitional attention control. He is still self-centered, but his perspective is influenced by a growing understanding of what is going on in the present situation. For instance, it becomes easier to accept that you have to agree on rules, and that many activities or games only work if everyone follows the same rules. This sets up an ongoing inner conflict between selfishly looking out for your own needs versus focusing on the group and your role in it. The child learns to see himself from outside; he is now able to compare himself to others, but he can also compare himself to his own and others' expectations. He also begins to realize that he has a wide repertoire of behaviors, not all of them appropriate.

Gradually, you develop many different sides to your personality …
a superhero …
a clown …
a mischief-maker …

… and a side that reflects on all the rest.

The child's ability to see himself from outside makes him care about what others think of him and his family. Being teased or mocked by other children, especially from his in-group, often triggers intense feelings of shame and embarrassment. His new external point of view makes it paramount for him to fit into the group. The good news is that he learns to balance his behavior towards others; the bad news is that he may go too far in adopting group norms. Driven to look and act exactly like the other members of the group, he extends this conformity to norms and morals. Anything that the group thinks and does is considered good and desirable.

With all these changes comes a tremendous development of the child's sense of self. His unique identity becomes more apparent, and he begins to be able to describe himself with reference to internal qualities, such as "kind" or "stubborn". He may also define himself through relationships, such as belonging to a certain school or sports team, or through favorite activities. He develops a much deeper perspective on others and is able to pick up not only on another child's immediate experience but also on the child's general life situation. As a result, he may develop impulses to help others, such as orphans, sick people or others in need.

Brain development

Between 6 and 9 years, a lot is going on in the brain. Primary neural networks in the subcortical areas, that control autonomic regulation and the organization of sensations develop further, and networks connecting these areas to the frontoparietal circuits expand. This, in turn, improves attention, motor control, executive function and prefrontal mentalizing processes. This sprawling network is also linked to emotional processes in the paralimbic cortex, which is associated with emotional processing, goal-setting, motivation and self-control.

There is a lot going on here, and despite the rapid growth of the child's mentalizing capacity, cognition and language skills, he probably still struggles with impulse control. By 10–12 years, the growth spurt has reached the frontal lobes and its connections with the rest of the brain, which leads to another tremendous increase in logical thinking, sense of time, grasp of cause and effect and the ability to plan. Parents often marvel at how sensible and grown up the child suddenly seems.

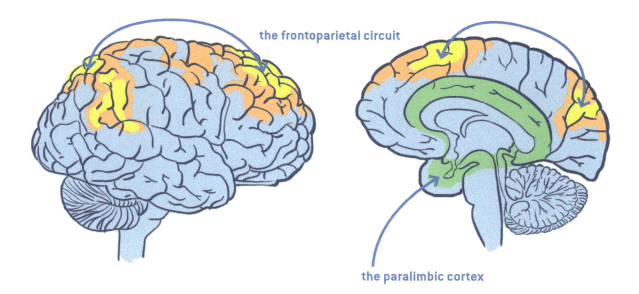

the frontoparietal circuit

the paralimbic cortex

From 6 to 9 years of age

Attachment relationships expand considerably during this phase. Many children find a "best friend" at this age, often – but not always – of the same sex. The ability to form and maintain good friendships remains important throughout life. Friendships take different forms for boys and girls, however. Girls form attachments with other girls by confiding in them and sharing secrets, while boys form attachments through competitive and often very physical play. Mixed-gender friendships often combine the two forms. In addition to a best friend or two, the child is often part of a small peer grouping in a larger, looser play group. The child often also joins other group contexts, such as a school class, a sports team and so forth. Along with his primary attachment to his family, these different attachment contexts place great demands on his skill at smooth code-switching, so that he can behave appropriately in the given context. It is important to be able to adapt to the social codes of the class or the sports team in order to be accepted. These codes may be in stark contrast to the codes in the family, and neither the peer group nor the family will normally put up with "inappropriate" behavior. Children from other cultures in particular struggle with colliding world views and have trouble determining who they really are and where they belong.

Around the age of 6 or 7 years, children's groups begin to establish well-defined sub-groups with specific norms and roles. Children become more conscious of their individual status among their peers, and as they increasingly compare themselves to others, their standing in important groups and networks becomes important for their sense of self-esteem. Status differences also affect the value of support or praise. If a student with a good singing voice sings in class and receives an admiring glance from a high-status classmate, his sense of self-esteem will grow. If the admiring look instead comes from a low-status kid, it will have less value and may even be unwelcome. Also, as mentioned in Chapter 4, status is expressed differently in all-boy groups compared to all-girl groups.

Boys are more likely to use commands, such as "Give me the ball!" and "Go over there!" ...

... while girls rely more on suggestions: "How about we play house instead?" or "Let's stop after your turn."

Boys are more likely to play outdoors and to play in larger groups with a leader. The leader is in charge and tends to reject suggestions from other boys. High status is attained by giving orders that others follow and by sticking together. Boys can also increase their status by telling stories and jokes that make them the center of attention, by engaging in risky or forbidden behaviors or by defying or challenging adult authority figures. Boys often boast of their accomplishments and argue over who is best at what.

Boys can attain high status by boasting of daring adventures.

A low-status boy will often be pushed around, and vulnerable boys may be teased and bullied. Boys especially enjoy competitive games, such as soccer and basketball, and when their activities are not innately competitive, they often team up to add an element of competition. They play games with winners and losers, and they make up rules – which are then hotly debated. Typical boy's play often involves running and chasing in large groups. Because of their explicit status dynamics, they need adults to maintain authority with clarity and kindness, offer appropriate challenges and help them develop honorable behavior in their hierarchies.

Girls' status is not primarily affected by acts of daring or combativeness, so they usually avoid risk behaviors and conflicts with adults. They are also not expected to boast about their accomplishments or to show that they are better than others. Girls' hierarchies are organized around social acknowledgement and attachments – better known as popularity and intimacy. Girls' play tends to involve fairly calm activities based on turn-taking or imitation, where the element of competition plays a relatively minor role. Girls' play typically has few rules; everyone is equal and performs according to standard rules, as in hopscotch, instead of competing against each other directly, as in tennis.

Orders are rarely used in all-girl groups. Girls are more likely to use suggestions or requests and tend to speak on behalf of the group, as in "We don't want to play with you." Proclamations from high-status girls will generally be approved, while comments from low-status girls are likely to be ignored. Studies show that girls tend to organize in pairs of BFFs, "best friends forever", and these pairs then form larger social networks. In group play, girls try to include their BFF. If a girl is not popular, she will often be excluded. Consequently, girls monitor their friendship relations for subtle shifts in alliances and strive to be friends with popular girls. Girls compete through gossip, where being the first to know something is a source of status. They also express attachment and closeness by confiding in each other and sharing secrets. Consequently, girls, especially high-status girls, need to find a balance between using information as gossip to boost their status versus keeping a secret to preserve attachment relationships.

Girls derive status from popularity and sharing secrets, so they keep an eye on subtle shifts in the friendships in their peer group.

Since sociability is an important value in girl groups, girls tend to avoid confrontations and conflicts and try to appear nice and cooperative. If one girl is annoyed with another, she typically confides in a third instead of confronting the girl in question – unlike boys, who tend to be much more up-front. This indirect approach gives rise to a high degree of complexity in a girl group and is often described in negative terms, such as going behind the other girl's back. However, it can also be understood as an attempt to keep the peace and avoid conflict. Because of their complex social relationships, girls usually have a greater need for adult assistance to create an inclusive group without backstabbing.

At this age level, both boys and girls develop the ability to role-play in larger groups and begin to establish more complicated *play narratives* that can be expanded endlessly. These pretend universes are often inspired by stories or films, such as *Frost* or *The Lion King*, but may also revolve around creative versions of reality, such as playing with toy cars. The ability to read others has become more sophisticated and can be used for good as well as bad. It can be used to get something you want, but are not entitled to, or to stir up a conflict …

The improved ability to read others can be used for mischief …

… by tricking others into blaming each other …

… but it can also be used to mentalize. Mentalization can be activated through empathy or cognitive processes. An empathic response might be the spontaneous impulse to comfort someone who is upset, while cognitive mentalization could involve finding the best way to mediate in a conflict.

… but the ability can also be used to empathize with both parties in a conflict …

… and to find a win-win solution.

At this stage, the ability to "see others from inside" is so sophisticated that children are able to include what they know about another person's life when reflecting on his or her immediate emotions. The ability to "see oneself from outside" at this stage involves seeing yourself and your family as you think others see and judge you, which can lead to feelings of shame.

The shame associated with being seen as bad or inferior is a recurring theme in stories and films. The protagonist either does something foolish in an important situation or is sabotaged and mocked. In movies and fairy tales as well as in real life, this shame is first regulated – accepted and transcended – by a loyal friend or mentor who stands by the protagonist, believes in him and helps him recover his determination. In typical storylines, shame turns to pride when the protagonist prevails by virtue of his heroism, kind heart and skills. Often, he is even gracious and generous towards the people who mocked him, unless they are considered truly evil, in which case they get their just punishment!

A true hero can reach out to his defeated enemy.

From 9 to 12 years of age

Shared norms and expanded attention are the basis for developing a conscience. As the child develops the ability to see himself from outside, his ability to assess himself and compare himself to others also grows. Accordingly, doing well becomes important, and he becomes more focused on competition and performance, for example in sport or school subjects. These comparisons affect the child's *status*. Status may be defined differently in different groups, just as individual status in a given group may vary over time. However, status may also "stick", so that a low-status child who is excluded by his peers may continue to have low status even after changing to a different school.

In groups of 9–12-year-olds – and in all groups, for the rest of your life – all the social motivation systems are active virtually all the time. Viewed as a whole, we can compare the energy in a group with the energy in an orchestra. A group can play an endless repertoire of interpersonal "music", ranging from sweet harmony to a jarring cacophony. The motivation systems can be compared to emotional musical instruments, where attachment, status, gender identity, play, cooperation and mentalization produce different kinds of "music", their sounds mingling and flowing together in interactions. Most of the time, this group "music" is played by our subconscious. Although most people have some sense of their own conscious intentions, we all have unconscious motives too. Moreover, much of our behavior is unconscious, and often we are unaware of the effect our behavior has on others.

As his mentalization capacity improves, the child begins to notice this dynamic: Both his own and others' behavior sometimes has unintended consequences. For instance, if a high-status child taunts or mocks a low-status child, the latter is humiliated in the eyes of the group. This may unexpectedly create a situation where the group splits into two factions, the group siding with the high-status child and the group that steps up to protect the humiliated child, with frosty relations or open warfare between them. You can see this dynamic play out in almost identical fashion in adult workplaces and in private groups.

The status motive behind your impulse to taunt or mock someone is often unconscious …

… and often leads to bullying and clique warfare in the group.

During this phase, the child develops a more complex understanding of moral behavior and conscience. It is no longer enough to be the leader or the most popular kid, you should also be a good person. With the support of both adults and peers, the child begins to feel responsible for the effect of his actions, taking pride in good deeds and feeling guilty about more dubious acts. At 8–10 years, children begin to develop a conscience, that is, the ability to feel bad about things they did. This guilty conscience can now be triggered by the mere thought of doing harm to others. With these increasingly complex thought processes, children begin to see that social and behavioral codes are not laws of nature but are created by common agreement, based on shared moral codes.

Everyday life includes countless interactions with adults and children in many different roles. Often, conflicts serve as the trigger for reflecting on what happened in yourself and in the other person – hopefully leading you to mentalize and achieve greater insight into your own and others' emotions. You begin to see that sometimes you try to get attention or approval by doing something stupid or telling a lie – and you realize that others sometimes do, too. With your increasing mental complexity, you may even discover that you have several different "inner voices" with wildly diverging goals and attitudes.

With better mentalization skills, you begin to be able to balance your own needs and desires with those of others. For instance, instead of basking alone in the glory of a championship, you will seek to share the credit and the honor. Improved self-insight and realism also help you set goals in life and strive to achieve them. At this developmental level, you know that, regrettably, you are not a great swimmer or soccer player, but you also know that with practice, you might become one.

Even if you are very clumsy at soccer …

… you can get really good with practice.

You also learn that getting better at something follows a certain structure. First, you figure out what you need to learn in order to reach your goals, then you come up with a plan that will move you in that direction, and then you set your plan in motion. Let us say you dream about becoming a good drummer and playing in a band with a bass player and a guitarist. You might begin by identifying the people or band you want to approach, then make a plan and practice hard to become a better drummer. When you think you are good enough, you work up your courage and make contact …

With realistic self-assessment, rejection can inspire you to develop a strategy …

… to strive to achieve your goal …

… to one day become part of the scene you are hoping to join.

Insight is hard work. Thousands of times during childhood, the child will need attention and support from grown-ups, from other children and from their own more mature aspects in order to understand themselves and others. Mentalization continues to develop and become more sophisticated throughout life, but the foundation is in place at the age of 10–12 years. During these years, the more mature identity begins to emerge, enabling the child to see himself more objectively, reflect on social interactions and make decisions about what role he wants to have in a group or how to handle difficult situations in the future.

Once you have learned to mentalize, the skill becomes internalized – like riding a bicycle. We do not need to think about it. We spontaneously understand and predict the behavior of others and reflect on emotions and interactions. When you have reached that level, identity development is driven by mentalization, and you develop a clearer understanding of your own choices and identity.

There are things about yourself that you do not like much.

Exploring our own and others' group norms and realism

Skills acquired from the age of 6 to 12 years – like the skills from earlier phases – may be more or less well developed in older children or adults. As in the previous chapter, you can use the questions below to explore the zone of proximal development in yourself or someone else. Again, the questions are phrased to invite self-reflection; it is still important first to look at those areas in yourself that you focus on in others. If you want more depth, you can either examine a specific situation, keep a particular question in mind for a while or exchange thoughts with an insightful friend or coworker.

Attachment
> Do you have any close peer relationships?

> Are you part of a workplace team or a recreational group – or both?

Play and cooperation
> Can you describe the norms and ground rules of one of your groups?

> When joining a new group, how good are you at noticing how others behave, and how good are you at fitting in?

> How does the group usually make decisions?

> How does the group usually resolve conflicts?

Gender identity and status
> If you look at one of your groups, are the group norms primarily male or female, or are they a mix of the two?

> Who has high status, and who has low status?

> What types of behavior increase status in the group?

> What is your status, how did you get it, and how do you maintain it?

Mentalization
> Think of a conflict you were in, and try to describe the other person's experience; can you do that in such a way that the other would recognize their experience?

> Could you describe your conflict from the point of view of an impartial observer?

> Are you able to make plans for joining a group you would like to be a part of? Would you be able to adapt your plan to match the circumstances? Describe a situation where you did that. How do you feel when you think about it?

> Can you describe another person's understanding of a third person's inner world? Describe a situation where two other persons were having a conflict. What was A's experience of B's experience? What did A do that made you think that, and how did B respond to A's response?

As you consider your answers, is there anything that you would like to take a closer look at or where you would like to experiment with a different behavior?

CHAPTER 7
Group culture, sexual maturation and identity

Personality development from 12 to 17 years of age

It is pretty simple, really: You enter adolescence a child, and you come out an adult. Adolescence is a time of constant and comprehensive reconstruction and construction chaos. During this prolonged renovation of the adolescent personality, young people reach sexual maturity – fall in love – seek a boyfriend/girlfriend, idols, values, group belonging – find an older mentor. They draw boundaries in relation to their family, in more or less dramatic ways. They seek the company of peers and search for a deeper meaning – or the wildest drinking binge – or their true identity – or the vocation of their adult life. They test themselves and their personal boundaries – they make decisions – or avoid making them – or change their minds – about education, romantic relationships, family relations, morality, their own personality and everything else under the sun.

Teenagers can be sensible, focused and goal-oriented …

… but they can also switch off reason in the blink of an eye.

The adolescent construction chaos is a result of dramatic changes happening in the body and brain, and the ensuing profound mood swings sometimes make it difficult for teenagers to maintain a realistic and balanced relationship with their surroundings. Common sense may be switched off at a moment's notice, so everyday appointments, commitments and consideration for others are drowned out by the general emotional turmoil.

During adolescence, the social motivation systems fuse to form the adult personality. In particular, the mentalization capacity expands and with it the ability to keep multiple perspectives in mind. As a result, teenagers now begin to be able to reflect on the values they were raised with and the ones they encounter in school and among their peers. They are able to incorporate other people's perspectives into their dreams for the future, as well as a general perspective of the work and living options available in their society.

Brain development

The teenage years are a critical period of the brain's development. While the childhood years were characterized by extensive growth of new neural connections, the entire brain is now undergoing a transformation, as it trims down to the adult level. In this process, brain cells die off and many synaptic connections are severed through pruning, while other connections are expanded to form central pathways and hubs.

During the teenage years the brain trims down in a process that begins at the back of the head and gradually moves forward.

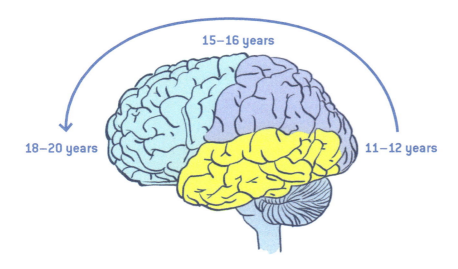

15–16 years

18–20 years

11–12 years

This cell death begins around the age of 12 years in the parietal lobe, which organizes our inner and outer sensations and our body identity. Around the age of 15–16 years it reaches the temporal lobes, which process emotions, sounds, visual impression and memories, before eventually – somewhere around the age of 16-18 years – reaching the frontal lobes. These areas are involved in all psychological processes, with special emphasis on organizing mentalization, planning, self-control, decision-making and executive functions. During the process of cell death and restructuring, function in the affected area is unstable – which is particularly evident in the teenager's intermittent loss of prefrontal reason and self-discipline.

Meanwhile, sex hormones are flooding the entire brain. As we mentioned in Chapter 4, the female sex hormones have a booster effect on *serotonin* and *oxytocin*, which have a calming influence and support impulses for calm, pleasant interactions with people you feel connected to. The male sex hormones, on the other hand, enhance the impact of neurotransmitters in the brain and body that lead to increased energy, risk-taking behavior and combativeness: *dopamine* and *adrenalin*. Dopamine is sometimes called the "brain's own speed". The brain's production of the neurotransmitter *GABA* (gamma-aminobutyric acid), which has a calming and sleep-inducing effect, is also increased. Thus, although teenagers are often full speed ahead when they are awake, they are also capable of sleeping in till late in the afternoon.

During this time *myelinization* also accelerates, a process where the major network connections are insulated with fat. This improves both the speed and the accuracy of brain functioning and enhances sensorimotor processes, emotions and cognitive skills – except, that is, when the brain is caught up in the construction chaos. The bridge between the two brain hemispheres, the corpus callosum, gets thicker, providing a closer connection between the stronger emotional expressions, concrete and holistic thinking that characterize the right brain hemisphere and the nuanced language skills and systematic approach of the left brain hemisphere.

The bridge that connects the right and left hemispheres, the corpus callosum, gets thicker.

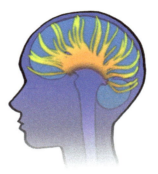

From 12 to 14 years of age

At this age, tweens and teens are strengthening their boundaries. They may put a sign saying: "NO ADULTS ALLOWED" or "PRIVATE – please knock before entering!" on the door to their room. They begin to distance themselves from their family members, are annoyed with their shortcomings and challenge parental boundaries and values. The major transformations raging through the teenager's body, hormone levels and brain also lead to intense mood swings. They search for their own separate identity and seek to reconcile their shifting sense of self with the dramatic changes that take place as their body develops, and they gradually reach sexual maturity.

Teenagers seek more attachments to people outside the family – often identifying strongly with a peer group. The issue of status in peer groups becomes more complicated, as various subcultures emerge and develop different values.

At this age, it is a disaster to be "different", and the experience of having high status in one context and low status in another can make it extremely difficult to stabilize the developing sense of identity. One boy may be a football star but struggle with cognitive challenges, while a skinny kid with no aptitude for sports may be admired as the local computer whiz.

In some groups, you can attain star status as a computer whiz …

… and in others, by showing off trendy clothes and a designer bike.

Some teenagers achieve high status because they are considered sexually attractive, while others are not yet sexually mature, and yet others are just "normal". Also, you may be excluded from a group because none of your qualities or skills are considered valuable in the dominant subcultures. Play impulses are now channeled into engaged *cooperation* in the core activities in the subculture, whether that happens to be computer games, sports, vandalism or wild parties. Sexual maturation and the first truly erotic infatuations or crushes occur at this age level. Emotional infatuation begins at the age of 3–4 years, but with sexual maturation comes a new degree of erotic tension, which adds new depths to the attachment to a boyfriend/girlfriend.

Sexual maturation expands attachment and gender identity by adding a strong sexual tension.

As the capacity for abstract thinking improves, their mentalization capacity expands yet again; but at the same time, the restructuring of the brain can switch off reason and logic without notice. So many changes are happening during this period that teenagers may lose the connection between past and present entirely; they are immersed in the moment without a thought for the consequences or prior experiences. Momentary emotional impulses can take over completely, and they are not aware that they are currently thinking, feeling and acting while their rational mind is in shutdown. When their reason switches back on, they may find it hard to fathom that they behaved the way they did; this often results in intense feelings of shame and guilt – and conflicts with authority figures.

Once reason is switched back on, teenagers often have trouble understanding how they could possibly have behaved in the way they did.

From 14 to 17 years of age

This period is characterized by intense passions and equally intense reflection. Teens strive for independence, trying in every way to mark their separation and distancing themselves from parents and other family members. They are intensely self-focused and fluctuate between having extremely high standards for their own looks and skills and feeling completely useless, like losers who will never amount to anything. To prop themselves up they hide behind a mask – an archetype.

Fortunately, we can prop up our fragile sense of self with a mask.

Paradoxically, while the teenager is busy rebelling and seeking independence, he still needs his parents' *attachment* and lifelong understanding of him to help him lower his defenses and develop a more loving and reasonable self-image. After a short while, however, the teenager once again feels the need to break free from the constrictions of family and childhood and dive back into his peer relationships. The erotic exploration from earlier years now develops into the first actual *romantic* and sexual relationships. Some of these turn into lifelong marriages or partnerships. Others last just a few months, as the two teenagers struggle to balance attachment patterns, self-image and sexuality, both within themselves and between them, in order for the relationship to survive and deepen.

The capacity for abstract thinking and complex *mentalization* improves steadily, although there will still be dramatic episodes, where rational thinking and self-control – the domains of the prefrontal cortex – are suspended, and the emotional limbic cortex pushes the teenager into sensation seeking and high-risk behavior. The intermittent shutdown of the prefrontal cortex also triggers the teenagers' intensely emotional perceptions of trivial interactions.

Teenagers seek out intense experiences together ...

The degree of risk-taking and sensation-seeking behavior depends on how the prefrontal cortex is working at any given moment. Impulsive behavior is increasingly kept in check by internalized moral principles at the prefrontal level. This helps the teenager distinguish between his own personal needs, some sense of the common good and conventional norms for "good behavior". When the prefrontal cortex is not switched off, it brings another huge expansion of the ability to reflect and mentalize, which can make it easier to resolve conflicts and to accept a level of personal responsibility for difficult interactions. Gradually, the teenager develops a firmer sense of his own identity, and increases his ability to imagine and set long-term goals.

From 17 to 20 years of age

Once again, mental perspectives expand. As teens develop a better sense of existential time frames, they may develop a passionate interest in the meaning of life and in spiritual or existential issues and values.

… but they also seek spiritual perspectives and search for the meaning of life.

The growing ability to grasp moral principles and the search for existential and spiritual meaning may lead to a deep commitment to universal concepts such as respect, equality and justice. At this age, young people are often attracted to political, environmental or charitable organizations.

With the broader perspectives comes the urge to fight for a better world.

Your concern for others increases, along with your awareness of the ways in which other people show care and concern. For instance, if you want to end a romantic relationship, you will worry about how to break the news. At this age, you also notice how others personify moral principles and kindness in difficult interactions with peers and teachers.

The changes in teens' brains and body are now proceeding at a calmer pace, which leads to greater emotional stability as well as a more secure sense of identity. That, in turn, leads to a greater sense of independence and self-reliance. As a result, they may have fewer conflicts in their families, and may also find it easier to find the courage to disagree with peers, even at the risk of jeopardizing their standing in the group. After a few years of being in opposition to the social and cultural traditions they grew up with, their sense of independence may now be strong enough that they can appreciate some of them. *Attachments* to peers and adults continue to develop, and they now establish friendships and group affiliations that can often last a lifetime. This deepening attachment also leads to a growing ability to share life's successes and failures, such as the disappointment of not being accepted into the school or program they dreamed of.

Peer bonds are now strong enough that you can rely on a friend for comfort when your hopes are dashed.

Play impulses continue to fuel teenagers' collaborations in leisure activities and educational projects. Many cognitive skills are stabilized during this phase, as they learn to reflect on complex issues, such as the relationship between their personal future and wider social and global dynamics. With these future perspectives, they develop an increased ability to delay gratification in favor of a later reward and a much better ability to plan a strategy. They are now able to reflect on issues and imagine the consequences of various paths of action. The observing self is now sufficiently stabilized that they can also explore inner experiences and narratives.

From early childhood, we seek to describe important things in words. Gradually, our narrative ability expands, and by the end of our teenage years, we are able to create a coherent biographical narrative, that is, a story about how we have become who we are. Through mentalization processes, our biographical narrative is modified repeatedly, during adolescence as well as later, in adult life, as new experiences come along and change our perspective on past events.

Exploring our own and others' group cultures, sexual maturation and identity

As in earlier stages, skills that develop during adolescence may be more or less well established. To determine your own or others' zone of proximal development, you can use the questions below for guidance. As in the previous chapters, the questions are phrased to invite you to mentalize yourself before mentalizing someone else. Once again, you should take a step further than simply thinking about the answer. You can either examine a real-life situation from your experience, reflect on one of the questions over a period of time or ask an insightful friend or coworker to consider the skill with you.

Attachment
> Who are the closest people in your life? What do you do when you are together?

> Are you a member of a group where you feel at home? What is it that makes it feel like home?

> How is your relationship with your original family?

> How do you engage with the larger world – what higher purposes or projects do you embrace or feel drawn to?

Play and cooperation
> What is your favorite way to "play"? What do you enjoy doing with others – and by yourself?

> How do you cooperate with others on these steps in project development: a) developing the idea, b) getting it off the ground, c) establishing fair and workable practical and economic terms, d) adjusting the project over time for optimal function, e) creating a good day-to-day structure for it and f) either completing it or passing it on to others?

Status
> Status perceptions are subjective. Our norms determine how we assess high and low status. What indicates high status and low status to you?

> Do you have someone that you consider a role model or a mentor?

Gender identity
> How does attachment, intimacy and sexuality combine in your life?

> How do you express your gender identity in behavior, clothing, leisure activities, choice of social circle and job choice? How happy are you with these aspects of your gender identity?

Mentalization
> How is your ability to understand and encompass different life events, so that you stay grounded in good times and don't give up during hard times?

> How is your ability to understand others and empathize with their life events, joys and sorrows?

> Do you pay attention to relationships in groups or families, and do you recognize patterns in the way people interact?

CHAPTER 8

Mentalization: the foundation for lifelong development

Development in adult life

In the previous chapters we have attempted to draw a map of personality development from childhood to early adulthood. We started out with a short description of some basic elements of this development, outlined five social motivation systems, and then devoted three chapters to the biologically and culturally determined unfolding of these systems.

Hopefully, it has become clear that personal maturation does not just happen with age. Fundamental skills need to be learned early in life. They, in turn, lay the foundation for more sophisticated skills to develop. If the fundamental social and personal skills are not stabilized, later developing skills are negatively affected, either becoming unstable or being used to overcompensate for a more fundamental deficiency.

From cradle to grave, personal maturation develops through interactions and life experiences with other people at the level of maturity that the personality and the nervous system are at. It is the daily dose of relevant connection, rather than any singular event, that makes the difference. Interactions that promote development must be embedded into the countless everyday exchanges and activities. Both children and adults need a relevant structure and goal, a living engagement, appropriate human care and flexible, relevant challenges. All of which, of course, is more easily said than done.

If the structure is too tight, it results in resignation and stress reactions ...

... if it is too loose, the engagement becomes weak and wavering.

Both can lead to frustration, harshness, stress reactions and mistakes …

… and finally, overcorrecting and defensive impulses may derail the work entirely.

The key is to create interactions where both tasks and needs are considered, and where group members can keep contributing to a well-balanced teamwork.

People are society's greatest resource. Our societies and our workplaces benefit when we improve our capacity for insight, at work as well as in our personal lives. When you are able to mentalize about yourself, you can integrate emotions and thoughts from different parts of your personality and work on your own development. This can help you to continue to refine your mentalization capacity throughout your adult life, as you go through life and are shaped by it: by love, conflicts and challenges in your intimate relationships with a life partner, with children and with work; as a citizen of a country and of the world and as a fellow human being; in your religious or spiritual endeavors; and in your inevitable meetings with loss, disease, suffering and death.

Embodied mentalization

Mentalization is about being able to embrace multiple perspectives. It is about being able to understand viewpoints and emotions and how they relate to behavior in yourself and others. To mentalize means to think and feel clearly, to relate to an experience with empathy and then take a step back and reflect on it. As we have seen, your emotional development can only unfold when your emotional experiences are deeply rooted in real-life experiences, in embodied sensations and in the ability to synchronize with others. Your cognitive grasp, too, depends on a connection with concrete and perceived reality. When you learned to count to three, you began by counting apples, balloons and so on. Gradually, those three apples and three balloons provided an inner sense of the symbol '3'. Together, your real-life skills in emotional relations and cognitive symbolization can eventually provide the basis for verbal mentalization. Thus, your sensorimotor experiences, which embed you in reality, your cognition and your emotions need to work together for you to keep developing your mentalizing skills throughout life. By examining how these three domains work together, you can discover developmental strengths and weaknesses in your adult mentalizing capacity. This can help you target specific skills from the previous chapters that you might wish to train – in a way that you can enjoy.

The mentalization triangle. A schematic image of the connection between sensorimotor, emotional and cognitive skills in personality.

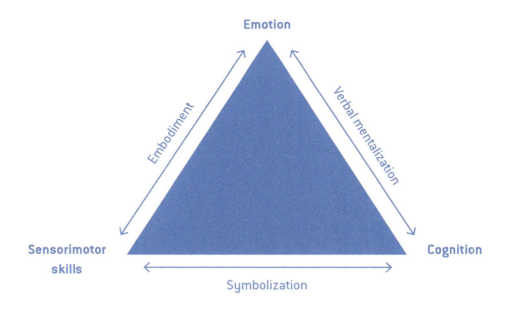

Embodiment

During our first years of life, emotional and sensorimotor development are interwoven and inseparable. They enable embodiment, personal grounding and emotional regulation. The secure child joyfully explores the world, supported by the parents' loving attention. The shared engagements and regulation in the actual process of exploring the world provide the child with the experiences that he will later learn to link to symbols; from apples and balloons to numbers.

Engaged and regulated meetings with others facilitate embodiment …

You will generally have developed a sense of self before you are 2 years old, largely based on the inner sensations of your emotional experiences in the countless interactions with parents and other adult caregivers. Due to the emphasis on thinking that characterizes the Western world, many people have forgotten this connection between embodied experience and emotions, but fortunately, you can recover it through practice. The bullet points below can give you some of the key elements, so you can explore the state of these skills in your everyday life and perhaps increase their integration in your mentalization processes.

The key elements of the embodiment side of the mentalization triangle are:

❯ Paying attention continuously to the connections between your inner sensations, facial expressions, movement impulses and all types of emotional states in yourself and others.

❯ Being able to be regulated, that is, soothed or encouraged, by emotional contact with others.

❯ Being able to lose yourself in a common project and to alternate between leading and following.

> Noticing when you are synchronized and seeking it, for example by spontaneously falling into step with someone or moving towards a sense of shared "flow" or shared meaning – those moments of meeting in a conversation that stand out in memory.

If you would like to read more about these skills, they are described in greater detail in *The Neuroaffective Picture Book* (2015, 2018).

Symbolization

Over the next few years, the child's emotional learning zone gradually shifts from the connection between body sensations and emotions to symbol use and pretend play, which is where emotions and cognitive skills meet. Words, for instance, are a type of symbols. A preschooler has not developed a sufficient grasp of language to describe emotional insights, so instead, he expresses his experiences through pretend play and symbols, such as using a doll as an airplane or a box as a castle. For pretend play, you have to have inner sensations and feelings for things, people or situations, and you must be able to transfer these meanings onto symbols and symbolic roles. These skills form the core of your ability to transfer information from one context to another; in emotional maturation, that means information about feelings and relationships. If you cannot express emotional states with symbols, your only remaining options are either to act out with emotional outbursts or impulsive behavior or "acting in" with self-judgement, depressed feelings or somatic symptoms. At the earliest stage of pretend play, the child might imagine that he is a brave hunter or a dangerous tiger. This kind of play is both embodied and symbolic. It coordinates sensorimotor experience and cognition, and it trains his embodied imagination, which in turn enables new emotional experiences. For instance, a favorite toy dog may become his wise and loyal friend and a comfort in difficult times; it may even have its own opinions and friends.

… embodiment, in turn, lets us create inner symbols to regulate our emotions …

... and so the symbols and qualities we embrace eventually become personality traits.

In your adult life, you have probably used symbols for cognitive purposes to achieve an overview, for example in sketching a new kitchen. However, you can also use them for emotional insight in a relationship by assigning symbols to different people or to different roles or parts of yourself and others. A classic psychotherapeutic model, Karpman's drama triangle, can help you to become more aware of three very common roles that are played out in human interactions: *victim – rescuer – perpetrator*. You might even recognize these three roles in your inner dialogues.

Karpman's drama triangle can play out as an inner dialogue or in interactions with others.

These are key elements of the symbolization side of the mentalization triangle:

> Being able to use symbols and roles to create an imaginary world where you can externalize thoughts and ideas, by yourself or with others.

> Being able to use symbols to explore relationships and interpersonal roles and to play with new possibilities.

> Being able to use symbols and roles to explore your experiences, internal states and aspects of your personality.

> Being able to transfer ideas from your pretend world to real life and from one context to another.

Mentalization

Embodiment and symbolization underpin the connection between emotions and thoughts, and that is the realm of the verbal mentalization capacity. Mentalization requires a development of cognitive skills that emerge during early school age. During childhood, you develop a better understanding of the relationships between inner experience and external behavior. It makes sense to you that a child can get upset over losing a game, then proceeds to get up and kick another child's toys, who in turn gets really angry and calls for a grown-up to come in and lay down the law. A child observing this sequence of events may feel empathy (embodiment) with both the other children and perhaps even with the adult. He may also use his imagination (symbolization) to reflect on what he would have done, or what they could have done differently. Using this sort of empathy and insight, the child learns to mentalize himself, other children and family members. He may also use it to follow complex interactions that involve many different people and emotions.

Your ability to mentalize renders other people's actions meaningful and predictable and this makes you more independent. It improves your ability to stay focused for longer periods, even if you are focusing on something boring but important, such as homework. Mentalization also helps foster *psychological intimacy*. This is the ability to sense and be present with the emotions and thoughts of another, without being overwhelmed by the other's suffering or discomfort. If your ability to achieve psychological intimacy is inadequately developed, you will either be overwhelmed by the pain of others or disengage from it.

These are key elements of the mentalization side of the mentalization tri-angle:

> Noticing and describing your own sensations, emotions, images, thoughts and actions and examining how they are connected.

> Imagining other people's sensations, emotions and thoughts, based on your observations and emotional empathy.

> Empathizing with yourself and others.

> Imagining how your behavior affects others.

Coordinating the three sides of the mentalization triangle
During adolescence, the physical, emotional and cognitive domains in-volved in mentalization undergo comprehensive development. However, mentalization is also shaped by the morals and the world view that young people absorb from their surroundings and by the identity and self-image they form. These years are crucial for the development of adult mentaliz-ing capacity. During earlier development phases, resources within each of the three domains may have developed, but during adolescence, the zone of proximal development involves balancing the three sides and weaving them together, so that you naturally – often without even thinking about it – activate relevant mentalization skills in your interactions.

These are key elements of the interweaving of the three sides of the men-talization triangle:

> Being able to sense your own personal blend of embodiment, assump-tions, expectations, observation and description and assessing whether there is too much or too little of each.

> Being able to balance embodiment, symbolization and verbal mentali-zation in relation to the situation you are reflecting on.

> Being able to let go of your own inner experience of a situation – or your empathy with another person – in order to reality-test your insight.

> Being able to see the situation in a broader context, such as in relation to others' needs or in relation to fundamental principles.

Closing remarks

In this book, we have described the complex unfolding of identity and socialization from around 2 years of age to the threshold of adulthood. Through countless interactions during childhood and adolescence, we have adapted to the culture that surrounds us. We have developed coping strategies for handling our emotions and staying organized under pressure, and hopefully we have formed realistic and positive expectations of the world around us.

Human beings are hyper-social, and relationships are crucial for our psychological development and well-being throughout the lifespan. We carry these relationships with us in the form of inner representations, and through them we are part of a much greater system, a society, which shapes us as well as the culture we live in. In turn, the tides of our society and culture constantly influence our personality and social relationships.

Society creates interactions and is created by them. If we are to build communities based on realism and efficiency, shared core values and the protection of vulnerable groups, children and young people need the guidance of adults – parents, school and preschool teachers and others – who are able both to mentalize about them and with them, and who can translate these insights into action.

In our globalized age, it is becoming increasingly clear that our behavior towards each other in our local communities affects dialogues and norms in the greater world. The anthropologist Margaret Mead once said: "Never doubt that a small group of thoughtful, committed citizens can change the world. Indeed, it is the only thing that ever has."

Thus, we end this book with the hope that it has offered inspiration for new, joy-filled and emotionally maturing interactions in the zone of proximal development – your own as well as that of others.

Recommended reading

If you are interested in more – or more theoretical – books about the neuroaffective perspective, you might try one of these books.

> Bentzen, M. (2015). *The neuroaffective picture book*. London: Paragon Publishing. Second edition, Berkeley, CA: North Atlantic Books, 2018.

This is the first picture book. It deals with basic personality development with the main focus on the development during the first two years of life. It contains descriptions of normal and stress-related interaction patterns in the triune brain and presents the neuroaffective compasses.

> Bentzen, M, Hart, S, (2015): *Windows of Opportunity – a neuroaffective approach to child psychotherapy*. London: Karnac Books.

This book uses the theory of neuroaffective developmental psychology and the model of the neuroaffective compasses to evaluate the relational and developmental factors in transcriptions of four child psychotherapies with four psychotherapists using different methods. The last chapters bring highlights from the discussions between the four therapists.

> Hart, S (2010): *The Impact of Attachment*. New York: Norton.

This book is one of the foundations of the neuroaffective personality development. It is a theory textbook compiling developmental research and the correlations to brain maturation.

CPSIA information can be obtained
at www.ICGtesting.com
Printed in the USA
BVHW02s1434250618
519969BV00011B/263/P